Guided Prayer Treatments

A Metaphysical Prayer Approach
to Receiving Your Blessings.

Karina G. Felix

Published by
Ingenious Publishers Inc.
West Palm Beach, Fl

Prayer Treatments for Health, Wealth, Peace and Joy.
A Metaphysical Prayer Approach to Receiving Your Blessings.
Copyright © 2023 by Karina G. Felix
All Rights Reserved. www.KarinaGFelix.com

Cover design by Karina Felix
Logo Design by Karina Felix
First Edition: January 2023
Printed in the United States of America

Felix, Karina G.
Metaphysics/ Prayer Treatments/Lifestyle
Print - ISBN: 978-1-7329696-5-0
Ebook - ISBN: 978-1-7329696-4-3
Library of Congress Control Number: 2022952223

Published and distributed by
Ingenious Publishers, Inc.
4805 Esedra Court
Lake Worth, Florida, 33467
www.IngeniousPublishers.com

No part of this book may be reproduced, scanned, or distributed in any print or electronic format without permission.

This book is designed to provide a guided prayer series to inform and inspire our audience. It is sold with the understanding that the publisher and the author are not engaged in rendering psychological, emotional, legal, or other professional advice. The content is a guide to support you and is the sole expression and opinion of the author, not necessarily the publisher. No warranties or guarantees are expressed or implied by the publisher's choice to include any of the content in this book. Neither the publisher nor the author shall be liable for any physical, psychological, emotional, financial, or commercial damages, including but not limited to special, incidental, consequential, or other damages. Our views and rights are the same: You are responsible for your own choices, actions, and results.

To request bulk order discounts and special sales purchases, please contact: Lifestyle@KarinaGFelix.com or Website: KarinaGFelix.com

PRAYER TREATMEMTS

I dedicate this
Prayer Treatment Booklet
to my parents,

Tirso F. Felix

and

Gladys M. Felix Peterson

whom I choose to join as part of my
spirit tribe during this cycle.

I've been blessed with the two best
teachers and students
to guide me on this journey.

Their creative and mental intelligence is the
driving force behind my life's works,
achievements, and accomplishments.

I could not have asked for better Earth Guides.

For this I give thanks,

I let it be so and so it is!

Dr. Karina G. Felix Ph.D Msc.

I acknowledge and bless
those that came before me,
and influenced me in so many ways.

And to those coming after me,
to offer guidance on their journey
during their own life cycle.

. . . and so it is!

⸻ PRAYER TREATMEMTS ⸻

ACKNOWLEDGMENT

My gratitude and appreciation
goes out to the people that stood
by me throughout my journey.

THANK YOU!

TESTIMONIALS

This collection of Prayer Treatments provides a valuable resource for connecting with our greatest treasure - the God-Consciousness that is within us. We have been conditioned to not connect with our True Nature, and therefore our True Power, believing that we are helpless victims of life's circumstances. Changing this belief is the first step towards manifesting the life we are meant to live, fulfilling our true Soul Purpose. These Prayer Treatments gently guide us to accepting and fully using the Power that is our birthright.

~ **Nandini Gosine-Mayrhoo** M.Msc.
Freelance Writer, Metaphysician. Author, *Nandi & The Music of the Plants*.

"Wow! I can feel the energy and power from the moment I opened this book. Every prayer holds the promise of transformation to tap into the storehouse of blessings the Divine has bestowed upon us. Thank you Karina for guiding us so perfectly to pause daily with these Prayer Treatments."

~ **Nancy Matthews**
Author, *The One Philosophy*.
Founder of Women's Prosperity Network

Prayer is talking to God.
Meditation is listening to God.

My beautiful, cherished friend and soul sister, Karina Felix, has written this sacred book filled with devotional prayers for us all to enjoy for spiritual upliftment and as a guide to deepen our conversations and connections with the Divine. Thank you for putting this powerful creation out into the world! I'm so proud of you. I Love you!

~ **Candy Barrotti** - The Loving Lightworker.
Christ Consciousness Spiritual Guide, Energy Healer, professional vocalist and recording artist.

PRAYER TREATMEMTS

Prayers should be a
vital part of your daily life.

By Asking, You Receive.

You affirm that the end results and
goals are already done!

You are a world class manifester.

... And So It Is!

Dr. Karina G. Felix Ph.D Msc.

PRAYER TREATMEMTS

TABLE OF CONTENTS

Preface 13
Introduction 15
Why Prayer Treatments? 17

Prayer Treatments for:

    Positive Mental Attitude 21

    Health & Wellbeing 25

    Financial Health & Wellness 29

    Confidence & Certainty 33

    Taking Inspired Action 37

    Positive Self Image 41

    Sourcing Energy 45

    Higher Conscious Awareness 49

    Prosperity & Abundance 53

    New Beginnings 59

    An Ideal Partner 63

    Serenity & Ease 76

    Successful Living 71

About The Author 75

Dr. Karina G. Felix Ph.D Msc.

PRAYER TREATMEMTS

Preface

You are profound intuitive guidance has led you to this book in anticipation of connecting with a genuine spiritual partner to guide and assist you as you find yourself and connect with your Soul and the journey it must undertake while it navigates this earth in physical form as part of you.

Attachment and possessions to material things in our life are a part of our daily worries and suffering. Believing that having more will make things better or even bring happiness has been proven repeatedly that it is not so.

Money and comfort are humble and vital missions to pursue, but they should not be the main and complete focus of a fulfilled lifestyle.

Learning, growing, and expanding with new

knowledge and wisdom should be your greatest quest on this journey. Teaching and sharing this wisdom is the next.

The one and only thing we get to accomplish on earth and take with us into the afterlife is our elevated Soul's Consciousness.

Your true legacy should be the pursuit of your Soul's Mission and being a willing participant in the expansion of Humanity and Spirit Consciousness.

Connecting with your true purpose will bring about a conscious way of enjoying and living a balanced, fulfilled lifestyle unleashing all of your potentials to achieve health, wealth, happiness, security, and peace.

By living your true Soul's purpose, you automatically attract to you a better lifestyle filled with prosperity, comfort, abundance, and joy. . . .

. . . And So It Is!

PRAYER TREATMEMTS

Introduction

Welcome. I am grateful for your presence as you utilize these prayer treatments to achieve a higher level of consciousness.

This consciousness allows you to be one with the Universal God-mind, individualized in the God-Power-Presence that dwells within each of us.

In each prayer session, focus on creating a deeper connection with your higher consciousness, entering into a new realm of existence.

During each prayer treatment create a serene calm area space where you are comfortable and will not be disturbed as you delve into a deep state of prayer meditation for your personal health, wealth and joy.

Dr. Karina G. Felix Ph.D Msc.

PRAYER TREATMEMTS

WHY PRAYER TREATMENTS?

All thoughts are prayers being manifested within your conscious and unconscious mind, whether you are aware of it or not.

One's outer life is generally a reflection of one's inner thoughts. To the degree that you are filled with faith is the degree that God's Presence is active in manifesting your prayer. Praying is not asking but attuning yourself to God's Will.

When you feel like you require physical healing, financial increase, prosperity, or any changes necessary to living a conscious happy lifestyle, you must accept that God already knows your desires.

Source is always guiding you towards your true ultimate purpose of happiness, wellbeing, prosperity, and abundance.

Prayer is a vital part of your life. Know that by asking, you receive. In other words, you affirm your desired end results and goals before they manifest in your life.

When you pray by asking for what you don't have, you create a premise that you are missing something, creating a sense of lack.

If you communicate this lack to the universe, it will respond by sending you more lack. Better said, it will return to you what you put out there.

By changing your perspective and believing that you already possess your desires and wishes, the universe will accommodate you and manifest them into your experience.

You must believe that you already possess what you seek, whether it is health, relationships, money, or happiness. Christ said: "Your Father knoweth what you have need of before you ask."

Know that the universe already knows what you desire, it is now up to you to show faith by accepting that your wishes are already on their

PRAYER TREATMEMTS

way. "Prosperity and abundance are your birthrights."

Use these prayer treatments as guidance as you navigate throughout your daily life's routine with these prayer requests, affirmations, and faith that your health, wealth & wellbeing are already done, restored and manifested into your reality.

This Prayer Treatments Series will help you reinforce your faith and desires so that you may receive your wishes. Your prosperity, abundance, and well-being are your birthright, and you deserve to receive all the riches, happiness, and joy that are actualizing in your presence at this moment and in your future.

Each day choose one of these Prayer Treatments that you feel you need in this moment to help you through any issues, discomfort, or confusion happening now.

Also use them as part of your gratitude prayers. To feel joy and appreciation and to bring more blessings into your life.

In these prayers I'll use the words Universal Intelligence, Source, God, within the prayers. You may replace them with whatever resonates best for you.

Dr. Karina G. Felix Ph.D Msc.

PRAYER TREATMENT FOR POSITIVE MENTAL ATTITUDE

Your prosperity, abundance and wellbeing are your birthright and you are deserving of receiving all the riches, happiness and joy that is actualizing in your presence.

This prayer treatment will guide you into knowing and accepting, without a doubt, that you are deserving of all your wishes, dreams, and desires.

Your subconscious mind does not know the difference between good or bad, right or wrong, positive or negative. You must consciously pay attention to your words, thoughts, and actions.

Like attracts like. So too negative attracts more negative, and positive attracts more positive. Being consciously aware, attentive, and guarding your thoughts, words, and actions should be part of your everyday exercise.

Take this moment to turn away from all distractions so you can commune with your higher god-centered self during these precious moments.

~
Pause

Take three deep breaths and allow these words to flow freely.

~
Prayer

I redirect my focus away from the outer world and turn my attention to the God-Power presence that dwells within me, bringing peace, harmony, and ease into my daily life.

PRAYER TREATMEMTS

Together we are one: One mind, one body, one soul, one spirit. Through the power and presence of God within me, I declare that I am already healthy, prosperous, and joyous.

I think positively each day, allowing the flow of energy and welcoming all levels of richness and comfort that seeks to improve my life.

My mind is open and receptive to receiving intuitive guidance from the universal intelligence so that I may enjoy peace of mind, body, and soul.

I surrender all control of my ego mind and allow the radiant light of Source to fill me with confidence, certainty, and satisfaction.

With a proper conscious mental attitude, I unblock any and all words, thoughts, and feelings that are not conducive to my prosperity and abundance.

I strengthen my mind through daily meditation and affirmations so that I may always see the good in others and myself.

I am self-confident as I allow the positive power of my universal God-Mind to govern all of my affairs.

I give thanks and bless my bills, checks, and payments as I exponentially increase the flow of money to attract itself back to me.

My mind, body, soul, and spirit are open to your control in everything I do, speak, or think. My words are like incantations, and I use them positively so that they may continue to bless and prosper me.

I am aware of my thoughts. I make daily efforts to keep them vibrating at an elevated level.

Luminous, radiant light radiates throughout my entire body, satisfying my body, mind, and soul with positive high vibrating frequencies.

As I arrive at peace within myself, I rest assured that you control my life now and eternally.

For this I give thanks,
I let it be so and so it is!

PRAYER TREATMEMTS

PRAYER TREATMENT
FOR
HEALTH & WELLBEING

This prayer is for your personal health and wellness. Dedicate this moment to yourself. Honor yourself as you read these words out loud.

It is important that you are in a state of harmony with your inner self. Be at peace with yourself, your body, mind, and soul.

Find a comfortable, quiet space where you will not be disturbed for the next few minutes as you focus your attention and intention on this moment and commune with Source.

Today you get to focus and pray for your personal health and wellbeing. As you read, gently close your eyes between sentences, concentrate on your body and let the words resonate with your subconscious mind.

~Pause

Take three deep breaths and allow your body to relax and bring your mind to the present.

~Prayer

I turn away from the outer world that surrounds me so that I may enter the quiet solitude of my inner being—a world without beginning or end.

I send love, peace, and blessings to all parts of my body; every organ, every skin surface, every cell, and every nucleus within me.

I am whole.
I am healed.
I am healthy.
I am complete.

My mind, which is in direct communication with the God-Power presence within me, is already at

PRAYER TREATMEMTS

work healing me from the inside out.

From the top of my head, to the sole of my feet, to the inner bright light dwelling inside, filling all space and crevices within each cell of my body.

Healing . . . Adjusting . . . Restoring to the perfect natural rhythm intended by the great designer of the universe, expressing perfect health in my body.

If a particular part of my body needs healing at this moment, I send that bright, radiant light of healing energy to all active cells in that area: Readjusting the energy flow and the energy activity into perfect harmony, aligning with the universe, with nature, and with God.

And so I move with the creative process within me. As the creative process flows through me, I open my soul to that flow that affects all areas of my physical body and now enters all levels of my mind.

All levels of my mind exist as multi-dimensional beings bringing healing to any trauma, any karma, or anything else that exists within me that needs healing and energy adjustment.

Right here, right now, that light lives within me, in perfect harmony bringing me perfect health. And as I feel this now-mentally, physically, and spiritually, my soul is at peace.

My Soul is in a state of divine harmony. Around my body is an aura of pure white beaming Christ-like light, protecting my body from any disruption from exterior sources, leaving me healthy and whole.

I am as perfect in physical body and mind as the creative presence of the universe dwelling within me.

I am still.
I am present.
I am one with the universal god-mind.

For this realization and acceptance of healing, health, wholeness, and all levels of my beingness,

> For this I give thanks,
> I let it be so and so it is!

PRAYER TREATMEMTS

PRAYER TREATMENT FOR FINANCIAL HEALTH & WELLNESS

In this session, you will pray for your financial health and security.

Your financial health and wellbeing are just as important as all other areas of your life. It allows you to enjoy a harmonious balance within all aspects of your daily living.

Begin as soon as you find a comfortable and quiet space where you will not be disturbed for the next few minutes so that you can focus this moment on just you.

This is part of your personal wellness and abundance treatment. At this moment, you will enter into a field of higher consciousness that you may commune directly with the Universal Intelligence.

As you settle into this quiet moment within a light meditational state, focus on your breathing, absorb the words, engage your emotions and relax into this prayer.

~ Pause

Take three deep breaths and allow these words to flow through you.

~ Prayer

My mind is part of the infinite mind of God, which contains the entire Universe within itself, therefore, I think of myself as a prosperous person.

My thoughts open channels, which allows whatever I need to flow to me.

PRAYER TREATMEMTS

I know that I do not have to dwell on any uncertainty because Source already knows what I need and desire.

I declare before the God of my being that God is my only trustworthy source of prosperity.

You know, God, what I have need of even before I ask. You are the ruling power of the Universe. I accept you as the power over all my financial affairs. Lead me, direct me, for I am open to receiving prosperity and abundance into my life right now.

Through the mystical power of Universal Intelligence working through me, I become intuitively aware of the positive success opportunities manifesting into my reality.

My prosperity includes love, health, wisdom, and happiness. I accept and receive the abundance that life has to offer. I know that there is more than enough for each and everyone one of us that walk on this earth.

I need not concern myself with that which is not mine because what belongs to me is already on its way to manifesting into my presence. I have no need for impatience, envy, judgments, or fear of scarcity. I am fully blessed with the ease of money and financial security flowing towards me.

My mind is always open to new ideas, inspirations, and intuition flowing from the Higher God-mind within me.

Each day I share in the wealth of God's abundance around me. I think positively each day of high-frequency energy vibrations welcoming all levels of abundance and prosperity entering into and improving my life.

I attune that the God-presence power within me is working daily in every area of my life, bringing me success and happiness.

My mind is open and receptive to intuitive guidance from Source. I am grateful for this moment and accept the fact that my wishes have already manifested into my existence.

For this realization and acceptance of abundance, prosperity, and well-being,

I give thanks,
I let it be so and so it is!

PRAYER TREATMEMTS

PRAYER TREATMENT FOR CONFIDENCE & CERTAINTY

During confusing and doubtful times, it is vital to turn to prayer to guide you as you navigate these uncertain times.

Find a safe and comfortable space as you focus within and dedicate this moment to yourself.

~
Pause

Take three deep breaths and feel your body relax and achieve a state of ease.

~ Prayer

I turn inward into my innermost God-Power presence within me, leaving the outer world behind for the moment.

I am aware of my oneness with the Universal God-mind. I trust in the God-presence within me that all outer situations taking place right now will not affect me because of my faith in the Universal Intelligence protecting my surroundings and me at this moment.

I do not live by fear but by faith, hope, and a deep understanding that I am the creator of my destiny.

The God in me can overcome any and everything. I do not allow outside events to affect my innermost sanctuary, which I diligently protect through prayers, my moments of meditation, and contemplation. I am enveloped by radiant, bright white light enfolding my aura, mind, and consciousness.

PRAYER TREATMEMTS

Inside of me, there is a divine plan for my Soul. I am one with the Infinite Intelligence and the Collective God-mind that connects me with everything and everyone within this Universe.

I have the capacity within the God part of me to get through whatever it is I must experience as part of my journey.

I have a firm base with my spiritual self-hood. I listen to what is happening in this world but am not influenced by it. I can see through the lies and deceit that are taking place around me and am thankful for the wisdom and knowledge open to me to not to accept it as my reality.

I turn away from all fear and uncertainty created by the outer world and listen to the still, small voice that emanates from within me.

Ultimately there is only one God, One Mind, One consciousness, expressed in various stages and cultures. I accept these differences we all have on this earth, and I judge no one for their beliefs and ideas, as I wish not to be judged in return.

I do not depend on external circumstances to bring me peace, ease, or happiness. I can only find that as I turn inwards to my innermost God-self.

The presence of God in me is eternally changeless, secure, whole, and complete.

My energy is uplifted, pure, and bright.
I am safe and at ease.

For these blessings,

I give thanks,
I let it be so and so it is!

PRAYER TREATMEMTS

PRAYER TREATMENT FOR TAKING INSPIRED ACTION

This prayer treatment is to help you overcome the roadblocks which keep you from taking action and receiving all the abundance, wealth, and rewards that are yours to enjoy.

This prayer treatment is essential because, in many instances, the small and mini actions not taken now will prevent the big rewards from arriving at your doorstep tomorrow.

Dr. Karina G. Felix Ph.D Msc.

Let's get started. Get into a comfortable, quiet place so you can place all your undivided attention on this prayer for your future well-being and success.

~
Pause

Take three slow deep breaths. Inhale deeply and slowly. Feel the ease course throughout your body and relaxing your mind.

~
Prayer

I have within me that extra something that can take me beyond self-doubt and uncertainty, that allows me to take a leap of faith and receive all of the riches and abundance that is due to me now.

I identify with my true self, my self that is one with Source. In so doing, I unfold the faith in myself that gives me the courage to take action.

I know I am never alone. I know that Christ-Consciousness is always with me and within me. I am constantly being guided and directed into living my best life.

PRAYER TREATMEMTS

Because of this inner knowing of a higher power at work, I am inspired to take action and make decisions in my daily life.

By allowing the ideas and thoughts that surface in my mind to manifest into my reality, I automatically have greater success, happiness, and fulfillment.

I am aware that the same higher mind that created the idea is also ready to help me succeed in manifesting it. I acknowledge that Source is the guiding force, and I practice continuous gratitude for its unfolding.

It is God's will that I live my best life; therefore, I have no need to doubt or ask how because I already know that the how is already taken care of. I have no need to figure it out on my own but to let it flow and let the Universal Intelligence orchestrate the manifestation of the idea into its reality.

I cancel all negative thoughts, doubts, and uncertainties from my mind. I replace them with success, determination, and self-confidence.

I am unique, as my purpose is specific just to

me, that I may express and share my gifts and wisdom with others. And so I continue on my true path to pure bliss, harmony, and blessings.

Life is good.
Life is great.
I am abundant and fulfilled.

*I give thanks,
I let it be so and so it is!*

PRAYER TREATMEMTS

PRAYER TREATMENT FOR POSITIVE SELF IMAGE

Today we will pray for your self-confidence, self-acceptance, and self-love. Ease into a comfortable chair where you can relax and focus on yourself.

~ Pause

Take three slow deep breaths, relax your thoughts and allow these words to flow over you.

~ Prayer

Each day my thinking and awareness can make or break my life and my potential. I am aware that the image I have of myself impresses itself upon my subconscious mind and can lead me to success and happiness or frustration and failure.

The image of myself that I put out there radiates out thoughts to others, which in turn attracts back to me the likeness of those thoughts.

I discharge positive emotional energy into the auric fields, which attracts positive people who can help advance my life towards success and abundance.

I am made in God's image; therefore, I can visualize and create whatever I see in my mind. I see myself as a healthy, prosperous, compassionate, intelligent, and loving person deserving of happiness, abundance, and love.

With every heartbeat, each day of my life, my subconscious mind accepts a visual picture of myself as a person of radiant health, financial wealth, fulfillment, and happiness.

Just as the whole Universal creation is within

PRAYER TREATMEMTS

the Universal God-mind, so too, I create and see in my personal mind the life I wish to live. I am a co-creator with Source working within me to manifest all my wishes, desires, and dreams.

Through the Grace and Power of this wisdom and blessings, I thank you, God, for all the good in my life, now and always. For this,

I give thanks,
I let it be so and so it is!

Dr. Karina G. Felix Ph.D Msc.

PRAYER TREATMEMTS

PRAYER TREATMENT FOR SOURCING ENERGY

Unbeknownst to ourselves, we allow others to pull energy out of us which leaves us drained, fatigued and confused. It is vital to replenish your energy consistently so that you don't fall into deep depression, anger and fears.

There are many ways to replenish your energy sources. You could meditate, journal, take a walk amongst trees, nature and beautiful scenes. Find a quiet place to contemplate or laugh for no good reason.

Give and receive love abundantly and enjoy the rewards of recharging your battery with free flowing high frequency vibrational energy.

~ Pause

Relax into a quiet comfortable chair or your bed to begin this prayer treatment. Take a deep slow breath and feel the bright radiant white light enter your lungs and course throughout your entire body sending positive healing energy.

~ Prayer

I acknowledge that for me to function at my maximum potential, I first must care, maintain and constantly refuel my energy supply.

I know that my body, mind and soul are as one unit and thus I must undergo frequent tune-ups for proper output of success and wellbeing in my life.

I am caring for this body that must last me throughout my lifetime, which I do by properly nurturing and nourishing it.

I nurture my inner success mechanism daily, so

PRAYER TREATMEMTS

it will take care of me throughout my life providing me with success, love, income, health and career.

I have the knowledge to know that my thoughts directly affect my body and my life, so I make all efforts to keep negative thoughts away from my mind.

I fuel up each day by meditating and resting to allow my body to regenerate, renew and heal. I give it proper rest, exercise and nourishment thereby generating a healthy mental attitude and success results.

I focus my mind on positive goal-oriented, spiritual self-image building thoughts. I establish in my consciousness that Source is unlimited in its ability to make the right things happen for me in my life.

I inhale peace and exhale negativity. I allow positiveness to enter my mind and push out all low vibrational energy thoughts that can affect my life in an unhealthy and unpleasant way.

I go into God-presence and allow it to refill my energy supply with high positive frequency healing and prosperous energy.

Dr. Karina G. Felix Ph.D Msc.

I give thanks for my success,

I let it be so and so it is!

PRAYER TREATMEMTS

PRAYER TREATMENT FOR HIGHER CONSCIOUS AWARENESS

Be grateful and present as you pray to achieve a higher level of awareness and consciousness.

This consciousness allows you to be one with the Universal God-mind, individualized in the God-Power-Presence that dwells within you.

In this prayer session, you will focus on creating a deeper connection with your higher consciousness and therefore enter into a new realm of your existence.

~Pause

First, find a serene, calm area where you are comfortable and will not be disturbed as you delve deeper into a light state of prayer meditation for your personal health, wealth, and happiness.

Take a deep, slow breath and feel the bright, radiant white light enter your lungs and course throughout your entire body, sending positive healing energy.

~Prayer

Entering a higher level of consciousness allows me to be a truly complete person, one who is inwardly aware and outwardly successful.

I accomplish this by acquiring a genuine contact with my true inner self-hood.

With a proper conscious mental attitude, I unblock any and all negative thoughts and passages that exist between my conscious and subconscious mind, which strengthens the expansion of my Inner Awareness and Higher Consciousness.

PRAYER TREATMEMTS

I focus primarily on positive and constructive mental attitude, negating and deleting all negative non-serving thoughts that may enter my mind.

Peace, understanding, and wisdom fill my entire being as I focus my mind on self-truth, the truth that I am one with Source.

My conscious mind is at peace as I go through this day because I am sensitive to the eternal peace of God's presence within me.

I am self-confident as I allow the positive power of my Universal God-mind to govern all of my affairs.

Anything that appears to go wrong this day is immediately corrected and made right by the unseen power of God working through me.

For all my blessings, good fortune, and eager anticipation of more good coming to me,

I give thanks,
I let it be so and so it is!

Dr. Karina G. Felix Ph.D Msc.

PRAYER TREATMEMTS

PRAYER TREATMENT FOR PROSPERITY & ABUNDANCE

Welcome to the wondrous way of living each day of your life. A way of life that will allow many of your dreams to become a reality.

~
Pause

Breathe in this bright, radiant white light and let it fill your inner being as you release all resistance and set the intention to receive plentifully and abundantly.

~Prayer

I take this moment to still my mind and free it from outer, daily-life thinking, and become aware of the internal consciousness of my mind.

My personal selfhood vanishes and is replaced with a sense of Universal essence, oneness with the Universe, and oneness with God.

I feel the inner light of wellbeing flowing through my body, mind, and soul.

I connect with the Ultimate Intelligence – the power and intelligence that can be manifested and demonstrated in my daily life.

As I take this moment to be still and sit in silence, it allows me to experience contact with higher intelligence.

I allow and am receptive to receiving intuiting and creative guidance from my higher mind.

I am prosperous already because I am ONE with the all-creating universal Source of all things

PRAYER TREATMEMTS

– the God within me.

I clear my subconscious mind of all negative thoughts so that I may enjoy maximum good health and abundance.

I now only allow positive thinking into my mind and reject all negative, self-destructive thoughts from entering my subconscious mind.

My conscious mind takes command of all incoming and outgoing thoughts that directly affects my environmental experience of daily living as well from subconscious recollections or influences.

As the ultimate mind or mystical power within me takes hold of my life, I will begin to experience numerous positive changes. It will slowly unfold into a better relationship with family and friends.

I feel more optimistic about myself and my environment and more in control of my life. I realize that higher intelligence is guiding my thoughts, therefore, also my life. I am opening up more opportunities both for my personal life and financial prosperity.

Dr. Karina G. Felix Ph.D Msc.

Through the mystical power of spirit working through me as I engage in daily activities, I become aware that I have reached my peak of peace and serenity.

I know that my body, mind, and soul are synced together, and I am now a walking manifestation of continual positive success in my life.

I understand that prosperity is a state of mind in which I experience happiness and peace within and about myself.

I live in continual expectancy, always open to new ideas, inspiration, and intuition flowing from my higher God-Mind into my experience.

I do not accept financial defeat. God is my only trustworthy source of supply and prosperity.

I am confident knowing that I possess a mystical magic, the inner presence of God working for my prosperity each and every day.

God is my true source of supply. I am already prosperous. I will shield myself from all negative influencers that are not conducive to my prosperity and wellbeing.

PRAYER TREATMEMTS

Having money is my birthright. I am surrounded by affluence and prosperity vibrations. God is already supplying me with money, health, and fun relationships in my life.

The higher god mind already knows my needs, which are already on their way to manifesting into my reality.

I take time each and every day to be in stillness, to commune with Source so that I may keep all channels open to receiving abundance, prosperity, and wellbeing.

I claim that prosperity is already mine. I will stay active and productive each day to allow an easy path for love, abundance, and ease to be a part of my ongoing journey.

For this I give thanks,
I let it be so and so it is!

Dr. Karina G. Felix Ph.D Msc.

PRAYER TREATMEMTS

PRAYER TREATMENT FOR NEW BEGINNINGS

Your best life is not an option; it's your birthright. You are deserving of all your wishes, desires, and more to come through. Always more, because as Souls having a human experience, its mission is to seek ongoing growth and expansion.

This is why humans must continue to self-educate and expand your knowledge-base and wisdom to assist your guides in bringing to you what you want to achieve and attain in this lifetime.

Seasons change. They show you that change is good and opens new possibilities and opportunities. Just as nature has its four seasons to showcase the beauty of new beginnings, so too do you have your own season changes taking place within you.

By observing and using these outer changes as your guidance, you have a chance to assess and redirect your life to get on the right path where you'll find the answers to all your questions.

Now let's pray that the new season of your life will bring you love, peace, harmony, laughter, great companionship, fun friendships, money in overflow, health in abundance, and peace of mind within your heart.

Pause

Get still and focused. Nestle into a comfortable, peaceful and quiet place. Take three slow deep breaths, hold on inhale for a count of three, then slowly exhale for a count and three, finish it with a smile on your face, and say out softly - "Thank you."

~Prayer

I take this moment to breathe in this pure, translucent, energy-filled air into my lungs, which then travels throughout my whole body.

I'm fully covered from the top of my head to my toes. I feel the lightness of healing, blessings, and protection entering into my beingness.

This energy flow is aligning and recharging my chakra energy centers as I dedicate this moment to receiving healing wellness energy coursing through my veins and replenishing all the energy centers within me with love and blessings.

I spend time seeking God's messages within myself. As I seek God, new beginnings begin to take place inevitably in my life.

I know what I know. I know that the conscious mind of my new beginnings is in perfect harmony and agreement with what the Universe has in store for me.

I am intuitively attuned to Source Energy, which will tell me when it's time to start something new or to act upon a decision that will propel me to be at the right time and place to be ready to receive my blessings.

I allow the Universe to create another masterpiece through me. I am open to receiving spontaneous inspiration that will lead me to take inspired action and receive the answers I seek to turn the pages to the following chapters of my life.

I already know who I am and accept myself just as I am. I am comfortable with my ideas and decisions. I am divinely guided to do what is suitable for myself right here, right now.

Heaven is within me. I establish and maintain contact with the ultimate ongoing creative consciousness to lead me to a state of peace, ease, and comfort.

This is the Power of Eternal new beginnings.

For this realization,

I give thanks,
I let it be so and so it is!

PRAYER TREATMEMTS

PRAYER TREATMENT FOR AN IDEAL PARTNER

You may have already met your ideal partner. They say we orbit around our desired partner long before we actually and eventually meet them in person.

You may have missed the opportunity to be in a loving, funfilled, devoted relationship with your ideal partner only because you have not stepped out into the world with an open mind and open heart.

Dr. Karina G. Felix Ph.D Msc.

Don't judge a book by its cover - meaning - don't always walk away from a potential match, because in the moment, it doesn' fit your initial ideas or expectations. Once you learn to see beauty all around you and hold space deep in your hearts for it, you will unblock and open the veils that are keeping your partner from showing up in your life and you from accepting them wholeheartedly within you.

Paũse

Be still and focused. Nestle into a comfortable, peaceful and quiet place. Take three slow deep breaths, hold on inhale for a count of three, then slowly exhale for a count of three, finish it with a smile on your face, and say out softly - "Thank you."

Praỹer

Dear God, I understand that you created everything, and as human beings, we were created in your image so that we may enjoy the beauty and appreciation of our surroundings.

Judging your creation is our human flaw. I turn to you to guide me away from this judgmental mentality flaw and lead me to acceptance and

PRAYER TREATMEMTS

openness to see beauty, kindness, and honesty everywhere, especially in unexpected places hidden from my blinded heart.

I choose to judge not from an unconscious state of mind. I request your guidance to minimize those traits from my psyche and allow me to open up to new fun surprises that you have in store for me.

My perfect partner is already within my reach. By taking the time to meet them first and getting to know their inner core beauty, I may pleasantly supprise myself by being open to see the true essence of their being, and thus my blessings becoming real.

I accept the perfect partner that you have reserved just for me. I am in the present space to allow them to truly see me, and I in turn see them through the eyes of our mutual Souls.

I judge not by looking at the outer exterior of my potenial partner but seek appreciation for their inner qualities. These inner qualities that bring out true love, respect, honor, and devotion. It is these qualities we get to love, honor and enjoy together.

Dr. Karina G. Felix Ph.D Msc.

This experience becomes beautiful because of the warm, shining, illuminating light flowing outwards from that person to me and vice versa from me to them. This inner brilliance is the true outer and inner beauty of our beingness.

This beauty I experience not through my eyes but through my heart; a knowingness that I'm in the presence of my one true soul and spirit partner; for now and eternity.

I shall focus on reaching the higher realm, the fifth dimension, that helps me see beauty like I've never experienced before.

I will sit with these feelings, allowing these words to vibrate through, in, and around my aura to feel the love, ease, peace, and blessings pouring over my head and coursing down over my entire body.

I feel the warmth of this energy force cascading around and inside of me. I sit still at this very moment to think, feel, and absorb these words and let them become an extension of Me.

I will be patient and persevere as I seek love and guidance from my higher intuitive God-pres-

PRAYER TREATMEMTS

ence to bring to me my spirit soul partner to share this journey with me

I am well - I am blessed - I am a child of the Lord.

> For this I give thanks,
> I let it be so and so it is!

Dr. Karina G. Felix Ph.D Msc.

PRAYER TREATMEMTS

PRAYER TREATMENT FOR SERENITY & EASE

This prayer is for serenity, ease, and peace of mind. This treatment allows you to come to a place to experience peace, tranquility, and comfort.

Your days are filled with ongoing to-do lists. Find time to do nothing. Time to find peace and quiet throughout your days.

Getting too caught up in your day-to-day schedule detracts you from your goals and delays your wishes that extends into days, weeks, and even months, and years when you realize that you have not given yourself a moment to just be.

Be in the present and be in sync with your inner journey. Not taking the time will eventually sneak up on you when you least expect it, and you will not be prepared and filled with the energy force that could help you get through moments of loss, separation, anxiety, doubts, and fear.

~ Pause

Let's begin by turning off all distractions and focusing on you. Yes, just you. This is your time, your moment to commune with your higher existence.

Feel the peace decending about you and creating a space of warmth and comfort. Inhale and exhale deeply three times. Bring your attention to the center of your mind.

~

PRAYER TREATMEMTS

Prayer

I am taking this moment to remind myself that I am unique, honored, blessed, and appreciated.

I maintain and establish control over my life, its success, and my happiness, by turning my personal mind over to the controlling Universal God-Mind.

This one Universal God-Mind has ultimate control over the outcome of all things in my life and experience.

The thoughts I think radiate out an energy field into the psychic atmosphere, attracting to me conditions that give me control over my success and happiness.

My finances are controlled and expanded by the power and wisdom of my Universal God-Mind.

Dear Divine presence of Universal GodMind, I turn inward to your presence at the center of my core.

My mind, body, soul, and spirit are open to your control in everything I do, think, and speak.

Peace and calm are established in me now through the knowledge that as I exist in your universal body, so too, you exist within my physical body.

In peace within myself, I rest in confidence that as you control the Universe, your power controls my life, now and eternally.

For your presence is vibrating success through me.

For this I give thanks, I let it be so and so it is!

PRAYER TREATMEMTS

PRAYER TREATMENT
FOR
SUCCESFUL LIVING

Welcome on this journey as you celebrate you taking action to live a successful, abundant lifestyle.

Living a succesful life takes just a shift of your mindset, allowing you to see things from a different perspective. This simple shift produces amazing manifestations of success, love, abundance and joy.

Dr. Karina G. Felix Ph.D Msc.

~
Pause

Being in a comfortable state of mind will enhance your prayer experience. As you prepare to get started, feel love pouring into your heart.

Breathe in deeply three time. Feel as the air enters and leaves your body. When you've reach a state of blissfulness ... start reading.

~
Prayer

I turn away from the outside world and turn inward to that place in my heart where serenity, peace, love, harmony, and abundance reside.

I am one with the universal God-Mind individualized in the God power presence within me.

Successful living is mine to the degree that I am aware and believe I am one with Source and affirm it is already mine.

By becoming self-aware, I recognize my personal relationship with the whole of life. I am intellectually and intuitively aware of how to guide my life and energy toward more successful living opportunities.

PRAYER TREATMEMTS

I have come from original Source and am one with the manifestation of that source. My mind, originating in the primal God-Energy center, is one with infinite wisdom every day of my life.

I express confidence in all I do every day. I live as an expression of the infinite unity, which vibrates through me as the energy of love, and through that vibration, attracts love back to me.

I direct these words to the infinite intelligence that dwells within me. I open my conscious mind and thoughts flowing through it each day to an ever-unfolding now greater self-reality with infinite presence.

I vibrate successful daily actions that steer me towards my dreams and desires.

I affirm this as the ultimate Source of success in my life, this moment and always.

For your inner presence vibrating success through me,

I give thanks,
I let it be so and so it is!

Dr. Karina G. Felix Ph.D Msc.

Listen to these Prayers Treatment online.

http://www.karinagfelix.com/prayer-treatments/

PRAYER TREATMEMTS

About

Karina G. Felix. is a Metaphysical Science Practitioner, Educator, and Ordained Minister.

She is the CEO of Meta Lifestyle Magazine, The Meta Lifestyle Magazine TV/Radio Show, Ingenious Publishers & Media Production and Visionaries in Motion, 501C3 non-profit organization.

She is a serial entrepreneur, magazine/book publisher, speaker, author, writer, 100 Successful Women in Business Award winner.

She has been on this spiritual journey since childhood, and understands the importance of passing all knowledge, and talents on to the next generation. Her life's journey is a testament to Living a Conscious Lifestyl on a daily basis.

Dr. Karina G. Felix Ph.D Msc.

My Mission

Karina is on a mission to support people on their spiritual journey and guide them to "Trusting the Process."

She wants to show people that they already have the answers within them and how they can uncover and understand the "What Now?" questions for deeper clarity.

"When you Remember your Purpose you will enhance and attract more feelings of love, ease, and prosperity in your life."

When you get to the core of YOU, your inner glow will light up the darkness within so that you can live a complete, fulfilled and balanced life.

Join me on a Guided Writing Prompt Series to unravel your doubts, fear, uncertainties and find happiness, prosperity and abundance.

Request more information

Website: www.KarinaGFelix.com
Email: Lifestyle@KarinaGFelix.com
Facebook: consciouslifestyleliving/
Twitter: LivingConsc825

Living The Conscious Lifestyle

Dr. Karina G. Felix Ph.D Msc.
Metaphysician

Everything you think you need to learn
to accomplish your goals,
You Already Know...... Now You Need To Apply It.

My Desire is to **Guide You** in **Achieving your Dreams and Desires** with **Simple Step-by-Step Principles** to get you to not only **Launch your Big Dreams** into **Physical Manifestations** but to lead you to **Create that Trust** in yourself to **Make that Leap!**

The Source for Enhancing, Impacting and Transforming Lifestyle

Read Inspiring Articles @
https://magazine.metalifestylemagazine.com/

Subscribe to
Meta Lifestyle Magazine (digital)
https://metalifestylemagazine.com/

Join us on Facebook
/MetaLifestyleMagazine/

Tune in to
The Meta Lifestyle Magazine Show
hosted by Karina Felix

YouTube: @MetaLifestyleMagazineShow
Facebook: TheMetaLifestyleMagazine/Show

Self-Healing | Self-Revelation | Self-Direction | Self-Guidance

INTUITIVE GUIDANCE
Using the Art of Journaling to Receive the Answers You're Seeking

Living The Conscious Lifestyle©   www.Lifestyle.KarinaGFelix.com

A Guided Writing Self-Enhancement Journey.

A Series of Deep Inner Analysis and Contemplation Exercises to guide you to attain your dreams, goals and desires..

Guided Writing Series

SELF-ASSESSMENT
SERIES # 1 - Aligning With Your True Self.

SELF-GUIDANCE
SERIES #2 - Creating Your Desired Lifestyle

SELF-REVELATION
SERIES # 3 - Becoming Self & Conquering Fear.

SELF-CONFIDENCE
SERIES # 4 - Finding Freedom and Confidence

SELF-HEALING
SERIES # 5 - Heal your Body, Mind & Soul

Each series comes with a 5-step guided prompt to get you to focus on what truly matters and how you can navigate any obstacles that present themselves to you.

Purchase the workbook
Email: Lifestyle@KarinaGFelix.com

www.ingramcontent.com/pod-product-compliance
Lightning Source LLC
Chambersburg PA
CBHW071253070526
44583CB00017B/2444